D1376294

Wiring Systems and Fault Fii
for Installation Electricians

Brian Scaddan

To my son, Stephen

Wiring Systems and Fault Finding for Installation Electricians

Brian Scaddan

NEWNES

Newnes
An imprint of Butterworth-Heinemann Ltd
Linacre House, Jordan Hill, Oxford OX2 8DP

$\mathcal{R}$ A member of the Reed Elsevier plc group

OXFORD LONDON BOSTON
MUNICH NEW DELHI SINGAPORE SYDNEY
TOKYO TORONTO WELLINGTON

First published 1991
Reprinted 1992, 1993, 1995

© Brian Scaddan 1991

British Library Cataloguing in Publication Data
Scaddan, Brian
  Wiring systems and fault finding for installation
  electricians.
  I. Title
  621.319

ISBN 0 7506 0072 1

Typeset by Vision Typesetting, Manchester
Printed and bound in Great Britain by
Biddles Ltd, Guildford and King's Lynn

Contents

Contents

3 Testing and test instruments 54

4 Fault finding 77

Preface

The aim of this book is to help the reader to approach the drawing and interpretation of electrical diagrams with confidence, to understand the principles of testing and to apply this knowledge to fault finding in electrical circuits.

The abundant diagrams with associated comments and explanations leading from the basic symbols and simple circuit and wiring diagrams, through more complex circuitry, to specific types of wiring system and, finally, to the methodical approach to fault finding.

Acknowledgement

I would like to thank Alan Burden for all his hard work in preparing the majority of the drawings and diagrams in this book.

1 *Diagrams*

BS 3939 symbols

British Standard 3939 gives the graphical symbols that should be used in all electrical/electronic diagrams or drawings. Since the symbols fall in line with the International Electrotechnical Commission (IEC) document 617, foreign diagrams *should* bear interpretation. Samples of the BS symbols used in this book are shown in Figure 1.

Kind of current and voltage

    —   Direct current

    ⌒⌣   Alternating current

    +   Positive polarity

    −   Negative polarity

Mechanical controls

    − − − − − −   Mechanical coupling

Figure 1   *BS 3939 symbols (continued on pages 2, 3 and 4)*

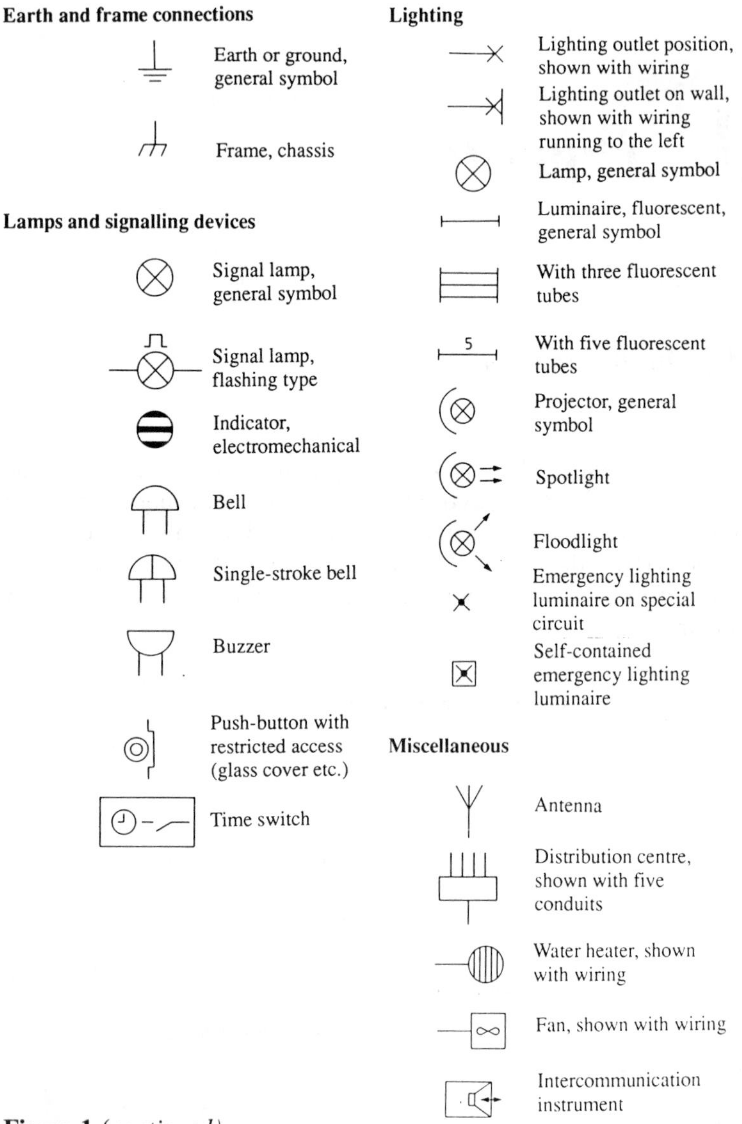

Earth and frame connections

Earth or ground,
general symbol

Frame, chassis

Lamps and signalling devices

Signal lamp,
general symbol

Signal lamp,
flashing type

Indicator,
electromechanical

Bell

Single-stroke bell

Buzzer

Push-button with
restricted access
(glass cover etc.)

Time switch

Lighting

Lighting outlet position,
shown with wiring

Lighting outlet on wall,
shown with wiring
running to the left

Lamp, general symbol

Luminaire, fluorescent,
general symbol

With three fluorescent
tubes

With five fluorescent
tubes

Projector, general
symbol

Spotlight

Floodlight

Emergency lighting
luminaire on special
circuit

Self-contained
emergency lighting
luminaire

Miscellaneous

Antenna

Distribution centre,
shown with five
conduits

Water heater, shown
with wiring

Fan, shown with wiring

Intercommunication
instrument

Figure 1 (*continued*)

Architectural and topographical installation plans and diagrams

Socket outlets

Socket outlet (power), general symbol

Three outlets shown: two forms

With single-pole switch

Socket outlet (power) with isolating transformer, for example shaver outlet

Socket outlet (telecommunications), general symbol

Designations are used to distinguish different types of outlet:

TP = telephone
M = microphone
⊏⫯ = loudspeaker
FM = frequency modulation
TV = television
TX = telex

Switches

Switch, general symbol

Switch with pilot light

Switch, two pole

Two-way switch, single pole

Intermediate switch

Dimmer

Pull-cord switch, single pole

Push-button

Push-button with indicator lamp

Switchgear, control gear and protective devices

Contacts

Make contact, normally open: also general symbol for a switch

Break contact

Change-over contact, break before make

Change-over contact, make before break

Make contact, early to close

Break contact, late to open

Make contact with spring return

Figure 1 *(continued)*

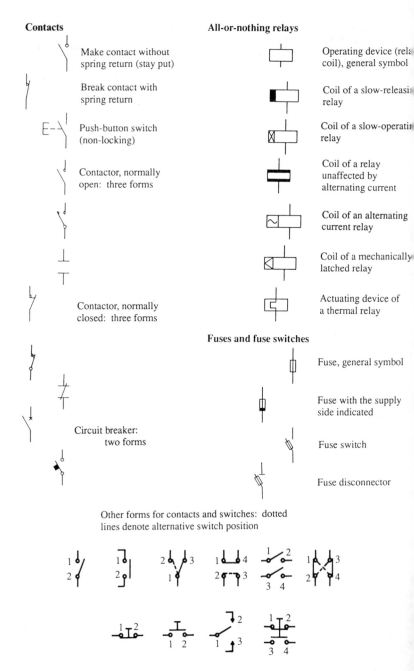

Figure 1 *(continued)*

Diagrams

The four most commonly used diagrams are the block diagram, the interconnection diagram, the circuit or schematic diagram, and the wiring or connection diagram.

Block diagrams

These diagrams indicate, by means of block symbols with suitable notes, the way in which an installation or system functions. They do not show detailed connections (Figures 2a and b).

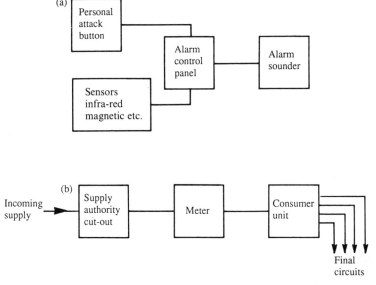

Figure 2 *(a) Security system. (b) Intake arrangement for domestic installation*

5

Interconnection diagrams

In this case, items of equipment may be shown in block form with details of how the items are connected together (Figure 3).

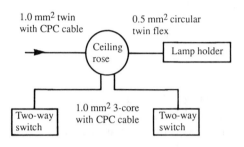

Figure 3 *Two-way lighting system*

Circuit or schematic diagrams

These diagrams indicate how a system works, and need pay no attention to the actual geographical layout of components or parts of components in that system. For example, a pair of contacts which form part of, say, a timer may appear in a different part of the diagram from the timer coil that actuates them. In this case, some form of reference scheme is needed, e.g. T for the timer coil and T1, T2, T3 etc. for its associated contacts.

It is usual for the sequence of events occurring in a system to be depicted in a circuit diagram from left to right and/or from top to bottom. For example, in Figure 4 nothing can operate until the main switch is closed, at which time the signal lamp comes on via the closed push-button contacts. When the push is operated the lamp goes out and the bell is energized by the push-button's top pair of contacts.

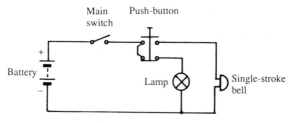

Figure 4

Wiring or connection diagrams

Here the diagrams show how a circuit is to be actually wired. Wherever possible, especially in the case of control panels, they should show components in their correct geographical locations.

The wiring between terminals may be shown separately or, when a diagram is complicated, shown in the form of one or more thick lines with terminating ends entering and leaving. Clearly, Figures 5a and b are the wiring diagrams associated with the circuit shown in Figure 4. Although Figure 5a would be simple to wire without reference to the circuit diagram, Figure 5b would present a problem as it is shown if Figure 4 was not available.

In either case an alphanumeric (A1, GY56, f7 etc.) reference system is highly desirable, not only for ease of initial wiring, but also for fault location or the addition of circuitry at a later date. Both circuit and wiring diagrams should be cross-referenced with such a system (Figures 6a–c).

Note how, in Figure 6c, at each termination the reference indicates the destination of that particular conductor. Also note how much more easily a circuit diagram makes the interpretation of the circuit's function.

7

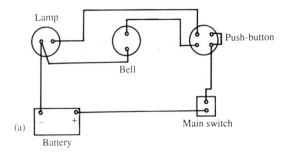

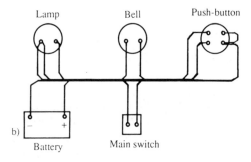

Figure 5

Circuit convention

It is probably sensible at this point to introduce the reader to circuit convention. This is simply a way of ensuring that diagrams are more easily interpreted, and is achieved by seeing that circuit diagrams are illustrated in a de-energized state known as *normal*.

Hence, if we take a new motor starter out of its box, all the coils, timers, overloads, contacts etc. are said to be in their normal position. Figures 7a–d illustrate this convention as applied to relays and contactors.

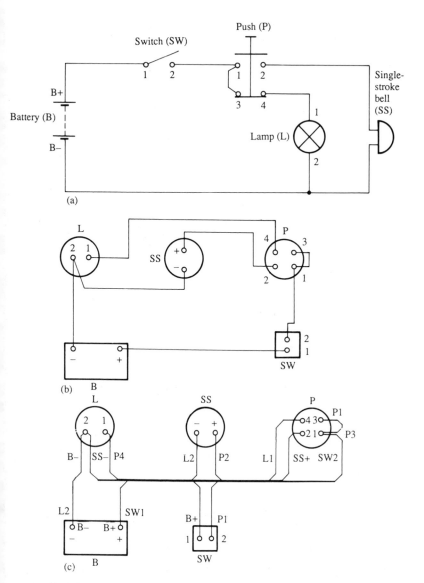

Figure 6

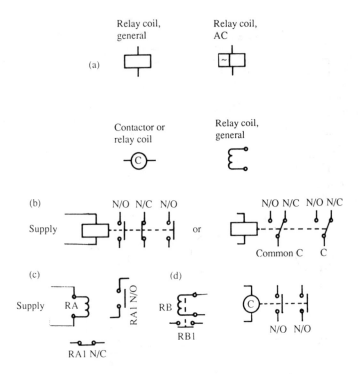

Figure 7 *Contactor and relay conventions*

Note that, provided diagrams follow this accepted convention, labelling contacts normally open (N/O) or normally closed (N/C) is unnecessary.

Constructing and interpreting circuit diagrams

In order to construct or interpret a circuit/schematic diagram of the controls of a particular system, it is necessary to understand, in broad principle, how the system functions. A logical approach is needed, and it may take the novice some while before all 'clicks' into place. Here is an example to consider.

Electronic valet

You work hard every day and return home late every evening. When you come in you look forward to a smooth Scotch, a sit-down and then a good soak in a hot bath. If you were acquainted with electrical control systems you could arrange for these little luxuries to be automated as shown in Figure 8.

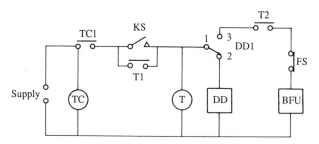

Figure 8

The system components in the diagram are as follows:

TC   Typical 24 hour time clock: TC1 is set to close at 2100 hours.

KS   Key switch operated by front door key: momentary action, contacts open when key is removed.

T   Timer which can be set to close and open contacts T1 and T2 as required.

DD   Drinks dispenser with a sprung glass platform. When energized, DD will dispense a drink into a glass. When the glass is removed, the platform springs up, closing contacts 1 and 3 on DD1.

FS   Normally closed float switch, which opens when the correct bath level is reached.

BFU   Bath filling unit: electrically operated hot water valve.

Let us follow the system through:

1   At 9.00 p.m. or 2100 hours the N/O contact TC1 on the time clock TC closes, giving supply to one side of the front door switch and timer contact T1.

2   You arrive home and open the door with the key, which closes the N/O spring-return contacts on the key switch KS, thus energizing the timer T. The drinks dispenser DD is also energized via its own N/C contacts DD1 (1 and 2).

3   The timer (now energized) instantly causes its own N/O contacts T1 to close, allowing supply to be maintained to T and DD via T1 (this is called a hold-on circuit) when you remove the key from the key switch. By the time you get to the lounge, the drinks dispenser has poured your Scotch.

4   When you remove the glass from the dispenser, DD1 contacts 1 and 2 open and 1 and 3 close, de-energizing the drinks dispenser and putting a supply to one side of the 10 minute timed N/O contacts T2.

5   You can now sit down, relax, and enjoy your drink, knowing that shortly contacts T2 will close and energize the bath filler unit BFU via the N/C water level float switch FS.

6   When the bath level is correct, the float switch FS opens and de-energizes BFU. You can now enjoy your bath.

7   One hour after arriving home, timer T will have completed its full cycle and reset, opening T1 and T2 and thus restoring the whole system to normal.

This system is, of course, very crude. It will work but it needs some sophisticating. What if you arrive home before 2100 hours: surely you need not stay dirty and thirsty? How do you take a bath during the

12

day without using the door key and having a drink? What about the bath water temperature? And so on. If you have already begun to think along these lines and can come up with simple solutions, then circuit/schematic diagrams should present no real problems to you.

Quiz controller

Here is another system to consider. Can you draw a circuit/schematic diagram for it? (A solution is given at the end of the book.)
  The system function is as follows:

1 Three contestants take part in a quiz show. Each has a push-button and an indicator lamp.
2 The quizmaster has a reset button that returns the system to normal.
3 When a contestant pushes his/her button, the corresponding lamp is energized, and energization of the other contestants' lamps is prevented.
4 The items of equipment required are: a source of supply; a reset button (push-to-break); three push-to-make buttons; three re-lays, each with one N/O and two N/C contacts; and three signal lamps.

The resulting diagram is a good illustration of the use of an alphanumeric system to show relays remote from their associated contacts.

Heating and ventilation system

Figure 9 is part of a much larger schematic of the controls for the heating and ventilation system in a large hotel. From the diagram it is

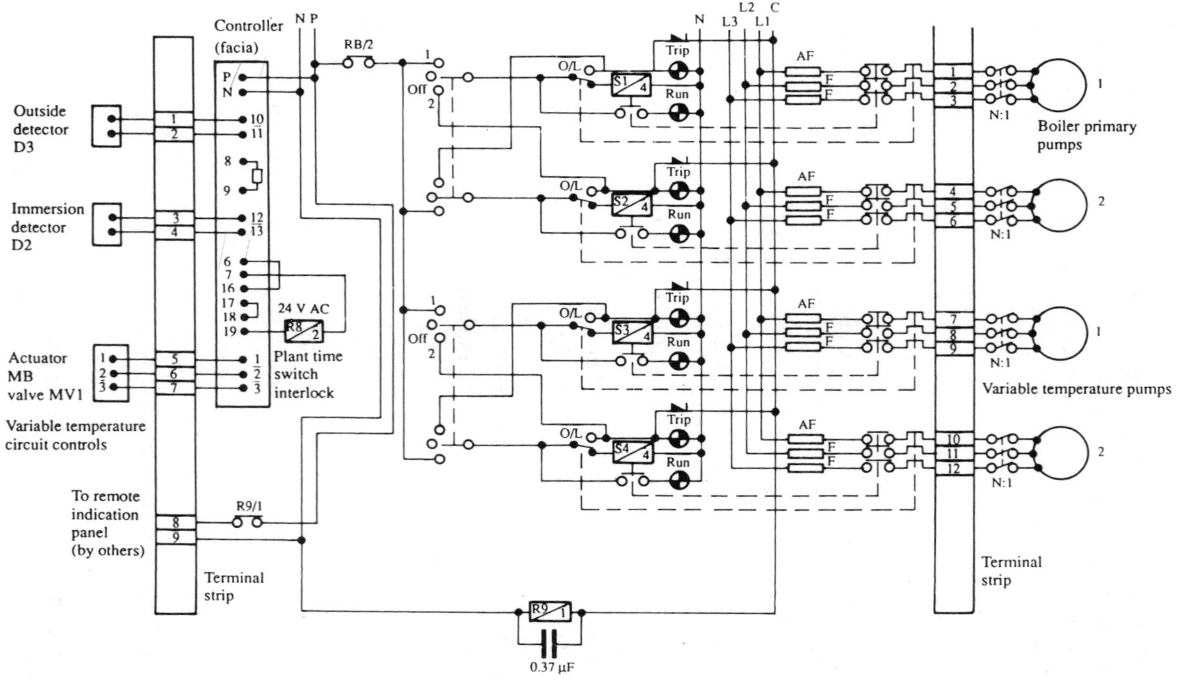

Outside detector D3
Immersion detector D2
Actuator MB valve MV1
Variable temperature circuit controls
To remote indication panel (by others)
Terminal strip
Controller (facia)
N P
RB/2
24 V AC
R8 2
Plant time switch interlock
R9/1
R9 1
0.37 μF
Off
O/L
S1 4
S2 4
S3 4
S4 4
Trip
Run
N
L3
L2
L1
C
AF
F
N:1
Boiler primary pumps
Variable temperature pumps
Terminal strip

Figure 9

relatively simple to trace the series of events that occur in this section of the system.

Clearly there are four pumps: two boiler pumps and two variable temperature pumps. One of each of these pairs is a standby in the event of failure of the other; this will become clear as we interpret the scheme.

There is a controller (similar to the programmer of a central heating system) which receives inputs from two temperature detectors and operates an actuator valve and a time switch. There are two sets of linked three-position switches, and direct-on-line three-phase starters with single-phase coils S1/4, S2/4, S3/4 and S4/4 for the pumps. There is also trip and run indication for each pump.

Let us now follow the sequence of events:

1 The selector switches are set to, say, position 1.
2 The temperature detectors operate and the controller actuates valve MV1. If the 24 V time switch R8/2 is energized then its N/O contact R8/2 is closed, giving supply to the selector switches.
3 Starters S1/4 and S3/4 are energized via their respective overload (O/L) contacts; the main contacts close and the pumps start. Auxiliary contacts on the starters energize the run lamps.
4 If the O/L on, say, pump 1 were to operate, then the N/O overload contacts would close, de-energize pump 1 and give supply to starter S2/4 for pump 2 via the second linked switch. At the same time the trip lamp would light, and a supply via a diode and control cable C would be given to relay R9/1, operating its N/O contacts R9/1 to indicate a pump failure at a remote panel. The diode prevents other trip lamps being back-fed via the control cable C from other circuits.
5 The reader will see that the same sequence of events would take place if the selector switches were at position 2 in the first place.

Relay logic

In the last few pages we have investigated the use of relays for control purposes. Whilst this is perfectly acceptable for small applications, their use in more complex systems is now being superseded by programmable logic controllers (PLCs). However, before we discuss these in a little more detail, it is probably best to begin with a look at relay logic.

We have already discussed circuit convention with regard to N/O and N/C contacts. In the world of logic, these contacts are referred to as 'gates'.

AND gates

If several N/O contacts are placed in series with, say, a lamp (Figure 10), it will be clear that contacts A *and* B *and* C must be closed in order for the lamp to light. These are known as AND gates.

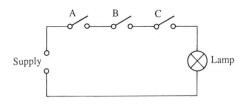

Figure 10

OR gates

If we now rewire these contacts in parallel (Figure 11), they are converted to OR gates in that contact A *or* B *or* C will operate the lamp.

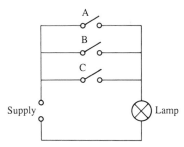

Figure 11

Combined gates

A combination of AND and OR systems is shown in Figure 12, and would be typical of, say, a remote start/stop control circuit for a motor. A *or* B *or* C will only operate the contactor coil if X *and* Y *and* Z are closed.

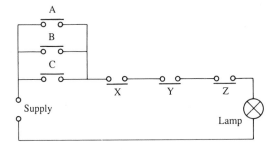

Figure 12

A simplification of any control system may be illustrated by a block diagram such as Figure 13, where the input may be achieved by the operation of a switch or sensor, the logic by relays, coils, timers etc., and the outputs in the form of lamps, heaters, sounders, contactors etc.

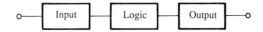

Figure 13

Programmable logic controllers

With complex control requirements the use of relays is somewhat cumbersome, and most modern systems employ PLCs. In basic terms these do no more than relays, i.e. they process the input information and activate a corresponding output. Their great advantage, however, is in the use of microelectronics to achieve the same end. The saving in space and the low failure rate (there are no moving parts) make them very desirable. A typical unit catering for, say, 20 inputs (I) and 20 outputs (O), referred to as a 40 I/O unit, would measure approximately 300 mm by 100 mm by 100 mm, and would also incorporate counters, timers, internal coils etc.

A PLC is programmed to function in a specified way by the use of a keyboard and a display screen. The information may be programmed directly into the PLC, or a chip known as an EPROM may be programmed remotely and then plugged into the PLC.

The programming method uses ladder logic. This employs certain symbols, examples of which are shown in Figure 14. These symbols appear on the screen as the ladder diagram is built up.

Here are some examples of the use of ladder logic.

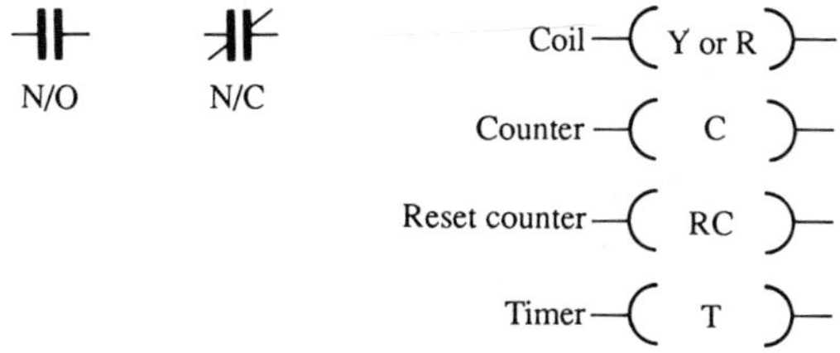

Diagrams

Figure 14

Motor control

Figure 15 illustrates a ladder logic diagram for a motor control circuit (no PLC involved here). Closing the N/O contacts X0 gives supply to the motor contactor coil Y0 via N/C stop buttons X1 and X2. Y0 is held on via its own N/O contact Y0 when X0 is released. The motor is stopped by releasing either X1 or X2.

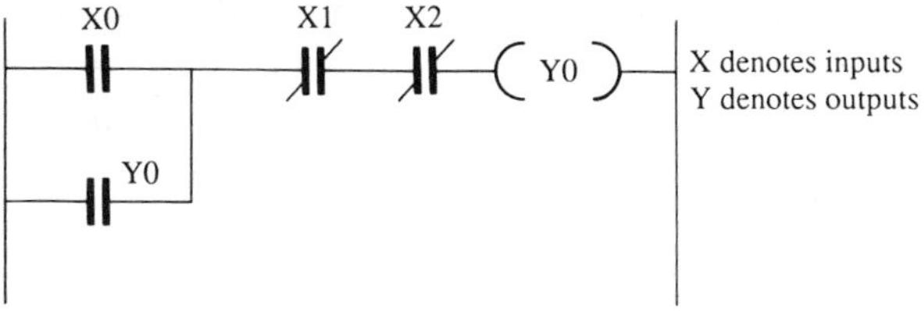

Figure 15

Packing control

Figure 16 shows the basic parts of a packing process. An issuing machine ejects rubber balls into a delivery tube and thence into boxes

19

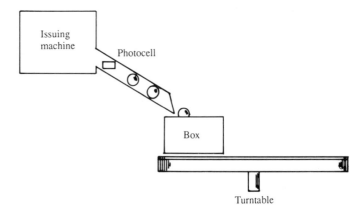

Figure 16

on a turntable. A photoswitch senses each ball as it passes. Each box holds ten balls and the turntable carries ten boxes.

Clearly, the issuing machine must be halted after the tenth ball, and time allowed for all the balls to reach their box before the turntable revolves to bring another box into place. When the tenth box has been filled, the system must halt and a warning light must be energized to indicate that the process for that batch is completed. When new boxes are in place the system is restarted by operating an N/C manual reset button.

This system is ideal for control by a PLC with its integral counters and timers. Figure 17 shows an example of the ladder logic for this system using the following:

X0 N/O photocell switch: closes as ball passes.
X1 N/C manual reset button.
Y0 Output supply to issuing machine.
Y1 Output supply to turntable.
Y2 Output supply to warning light.
C0 Internal counter set to 10 with one N/C and two N/O contacts.

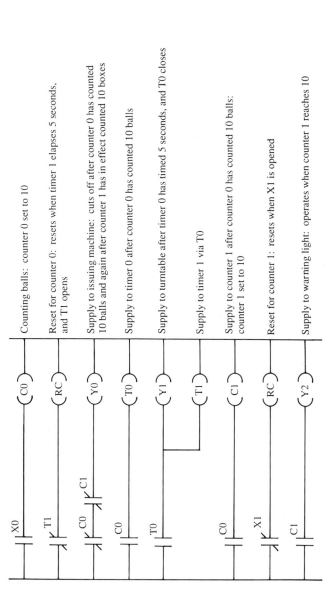

X0 ——(C0)	Counting balls: counter 0 set to 10
T1 ——(RC)	Reset for counter 0: resets when timer 1 elapses 5 seconds, and T1 opens
C0 —C1—(Y0)	Supply to issuing machine: cuts off after counter 0 has counted 10 balls and again after counter 1 has in effect counted 10 boxes
C0 ——(T0)	Supply to timer 0 after counter 0 has counted 10 balls
T0 ——(Y1)	Supply to turntable after timer 0 has timed 5 seconds, and T0 closes
——(T1)	Supply to timer 1 via T0
C0 ——(C1)	Supply to counter 1 after counter 0 has counted 10 balls: counter 1 set to 10
X1 ——(RC)	Reset for counter 1: resets when X1 is opened
C1 ——(Y2)	Supply to warning light: operates when counter 1 reaches 10

Figure 17

C1   Internal counter set to 10 with one N/C and one N/O contacts.
T0   Timer set for 5 seconds with one N/O contact.
T1   Timer set for 5 seconds with one N/C contact.
RC   Reset counter: resets a counter when supply to it is cut.

Fault location

Another major advantage of the use of PLCs for controlling systems is the relative ease of fault location. In the event of system failure, the keyboard and screen unit is plugged into the PLC and the condition of the system is displayed in ladder logic on the screen. Then, for example, any contact that is in the wrong position will show up.

Drawing exercises

1   Using BS 3939 architectural symbols, draw *block* diagrams of the following circuits:
    (a) A lighting circuit controlled by one switch, protected by a fuse, and comprising three tungsten filament lamp points, two double fluorescent luminaires, and one single fluorescent luminaire.
    (b) A lighting circuit controlled by two two-way switches, protected by a fuse, and comprising three floodlights.
    (c) A lighting circuit controlled by two two-way switches and one intermediate switch, protected by a circuit breaker, and comprising three spotlights. One of the two-way switches is to be cord operated.
    (d) A ring final circuit protected by a circuit breaker, and comprising six double switched socket outlets and two single switched socket outlets.
2   Replace the symbols shown in Figure 18 with the correct BS 3939 symbols.

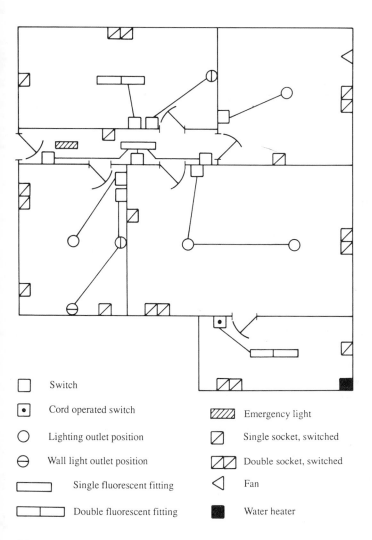

☐	Switch		
⊡	Cord operated switch	▨	Emergency light
○	Lighting outlet position	◨	Single socket, switched
⊖	Wall light outlet position	▧	Double socket, switched
▭	Single fluorescent fitting	◁	Fan
▭▭	Double fluorescent fitting	■	Water heater

Figure 18

Solutions are given at the end of the book.

2 *Wiring systems*

In this chapter we will investigate a selection of the many wiring systems employed in modern installations. Some of these systems are simple to understand and require little explanation. Others of a more complex nature should now, in the light of the reader's new-found knowledge of diagrams etc., present only minor problems of interpretation.

Radial systems

Any system which starts from the supply point and either radiates out like the spokes of a wheel, or extends from one point to another in the form of a chain, is a radial system. Figures 19, 20 and 21 illustrate such systems as applied to lighting and power circuits.

It should be noted that BS 3939 architectural symbols are not often shown in this fashion; it is usual to see them used in conjunction with building plans. This will be discussed later.

Ring circuits

These circuits start at the supply point, loop from point to point and return to the same terminals they started from. They are most

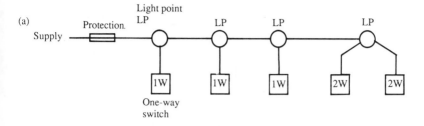

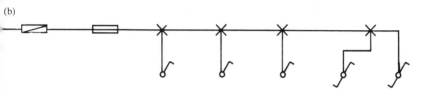

Figure 19 *Radial lighting circuit using (a) representative (b) architectural symbols*

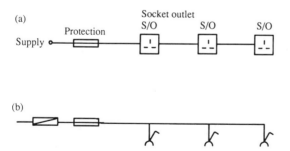

Figure 20 *Radial socket outlet circuit using (a) representative (b) architectural symbols*

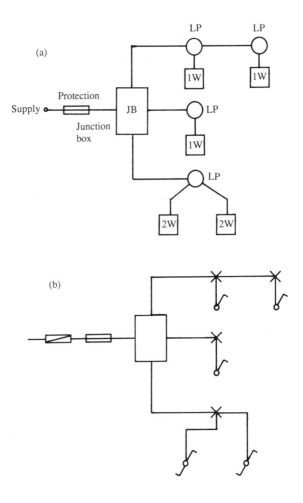

Figure 21 *Radial lighting circuit using (a) representative (b) architectural symbols*

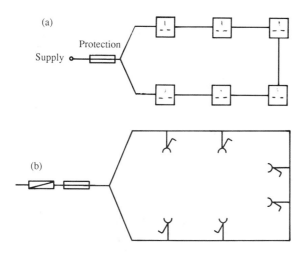

Figure 22 *Ring final circuit using (a) representative (b) architectural symbols*

popular in domestic premises, where they are referred to as ring final circuits. However, such systems are also used in factories where overhead busbar trunking is in the form of a ring, or for supply authority networks (Figures 22 and 23).

Distribution systems

Such systems are many and varied, but they are quite simple to understand as they tend to follow the ring and radial concepts.

Take for example the UK electricity system. Regardless of who owns this or that part of it, the system functions in the following stages: generation, transmission and distribution. Generated electricity is transmitted over vast distances around the UK in a combination

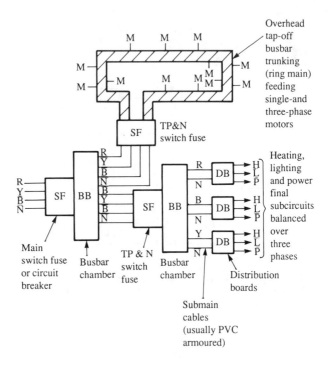

Figure 23 *Layout of industrial installation*

of ring and radial circuits to points of utilization, where it is purchased by the electricity boards and distributed to their customers. Once again these systems are in the ring or the radial forms.

Probably more familiar to the installation electrician is the distribution system in an industrial or commercial environment. Here one finds radial circuits originating from the intake position and feeding distribution boards (DBs), from which are fed either more DBs or final circuits. Diagrams for such systems may be of the block type (Figure 24) or of the interconnection type (Figure 25).

Note how much more detail there is on the section of the drawing shown in Figure 25. Cable sizes and types are shown, together with

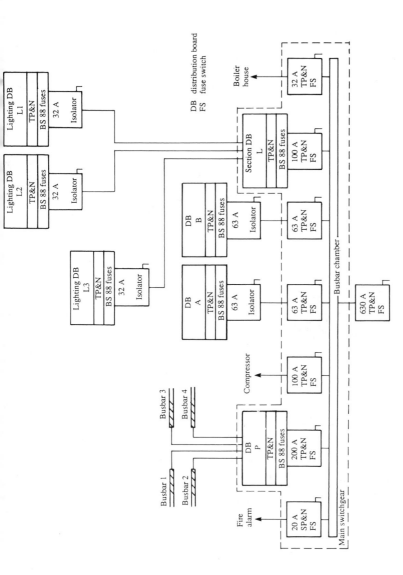

Figure 24 *Distribution system, block type*

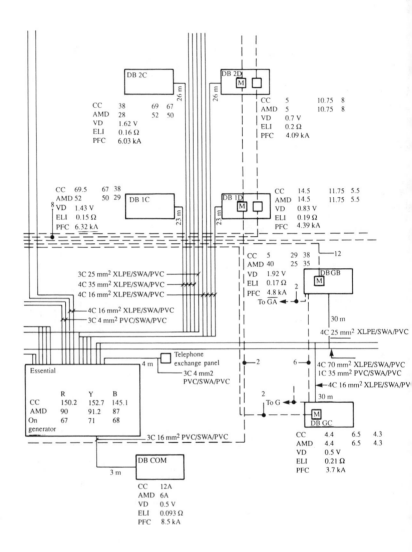

Figure 25 *Distribution system, interconnection type*

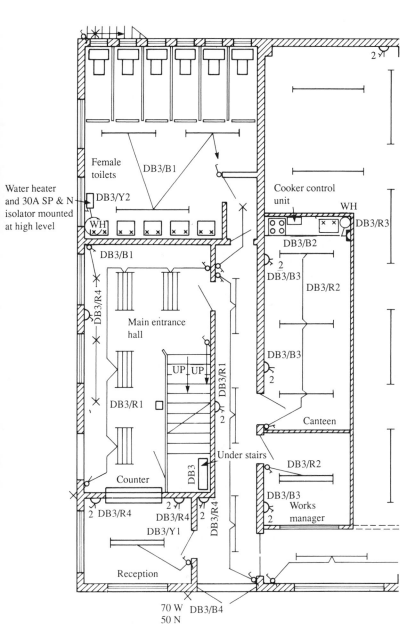

Figure 26 *Example floor plan*

cable lengths (23 m, 26 m etc.). Details at each DB indicate current loading (CC), approximate maximum demand (AMD), voltage drop (VD), earth loop impedance (ELI) and prospective fault current (PFC).

With the larger types of installation, an alphanumeric system is very useful for cross-reference between block diagrams and floor plans showing architectural symbols. Figure 26 shows such a system.

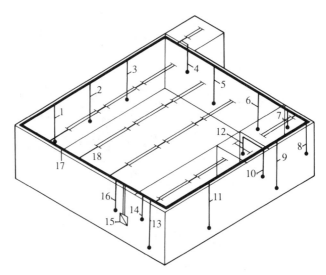

Figure 27 *Isometric drawing for garage/workshop*
1   *Three-phase supply to ramp: 20 mm² conduit*
2   *Single-phase supply to double sockets: 20 mm² conduit.*
    Also 3, 5, 6, 9, 11, 13
4   *Single-phase supply to light switch in store: 20 mm² conduit*
7   *Single-phase supply to light switch in compressor: 20 mm²*
    conduit
8   *Three-phase supply to compressor: 20 mm² conduit*
10 *Single-phase supply to heater in WC: 20 mm² conduit*
12 *Single-phase supply to light switch in WC: 20 mm² conduit*
14 *Single-phase supply to light switch in office: 20 mm² conduit*
15 *Main intake position*
16 *Single-phase supplies to switches for workshop lights: 20 mm²*
    conduit
17 *50 mm × 50 mm steel trunking*
18 *Supplies to fluorescent fittings: 20 mm² conduit*

Distribution board 3 (DB3) under the stairs would have appeared on a diagram such as Figure 25, with its final circuits indicated. The floor plan shows which circuits are fed from DB3, and the number and phase colour of the protection. For example, the fluorescent lighting in the main entrance hall is fed from fuse or MCB 1 on the red phase of DB3, and is therefore marked DB3/R1. Similarly, the water heater circuit in the female toilets is fed from fuse or MCB 2 on the yellow phase, i.e. DB3/Y2.

Figures 27, 28 and 29 illustrate a simple but complete scheme for a small garage/workshop. Figure 27 is an isometric drawing of the garage and the installation, from which direct measurements for materials may be taken. Figure 28 is the associated floor plan, which cross-references with the DB schedule and interconnection details shown on Figure 29.

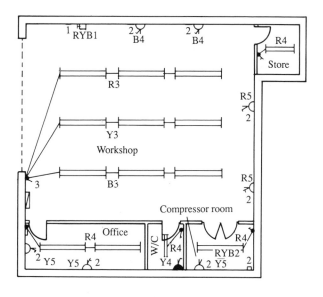

Figure 28   *Floor plan for garage/workshop*

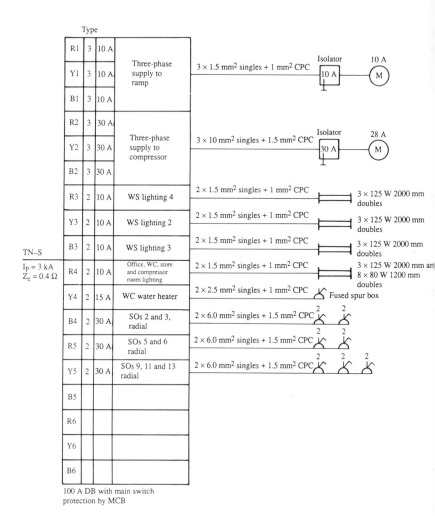

Type						
R1	3	10 A	Three-phase supply to ramp	3×1.5 mm² singles + 1 mm² CPC	Isolator 10 A	10 A (M)
Y1	3	10 A				
B1	3	10 A				
R2	3	30 A	Three-phase supply to compressor	3×10 mm² singles + 1.5 mm² CPC	Isolator 30 A	28 A (M)
Y2	3	30 A				
B2	3	30 A				
R3	2	10 A	WS lighting 4	2×1.5 mm² singles + 1 mm² CPC		3×125 W 2000 mm doubles
Y3	2	10 A	WS lighting 2	2×1.5 mm² singles + 1 mm² CPC		3×125 W 2000 mm doubles
B3	2	10 A	WS lighting 3	2×1.5 mm² singles + 1 mm² CPC		3×125 W 2000 mm doubles
R4	2	10 A	Office, WC, store and compressor room lighting	2×1.5 mm² singles + 1 mm² CPC		3×125 W 2000 mm and 8×80 W 1200 mm doubles
Y4	2	15 A	WC water heater	2×2.5 mm² singles + 1 mm² CPC		Fused spur box
B4	2	30 A	SOs 2 and 3, radial	2×6.0 mm² singles + 1.5 mm² CPC		
R5	2	30 A	SOs 5 and 6 radial	2×6.0 mm² singles + 1.5 mm² CPC		
Y5	2	30 A	SOs 9, 11 and 13 radial	2×6.0 mm² singles + 1.5 mm² CPC		
B5						
R6						
Y6						
B6						

TN–S
$I_P = 3$ kA
$Z_e = 0.4 \ \Omega$

100 A DB with main switch
protection by MCB

Figure 29 *Details of connection diagram for garage/workshop*

A similar diagram to Figure 29 is shown for part of the ventilation system in a commercial premises in Figure 30.

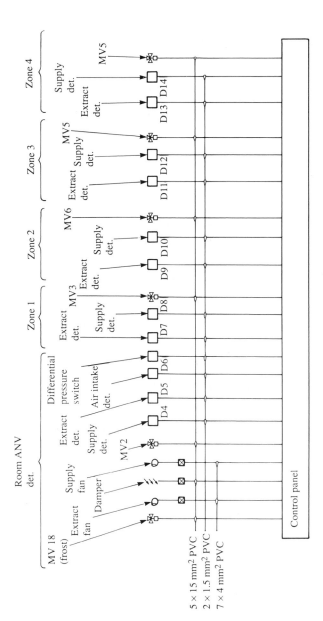

Figure 30 *Connection diagram for part of ventilation scheme (det.: detector)*

Emergency lighting systems

These fall into two categories: maintained and non-maintained. Both these systems may be utilized by individual units or by a centralized source.

Maintained system

In this system the emergency lighting unit is energized continuously via a step-down transformer, and in the event of a mains failure it remains illuminated via a battery (Figure 31).

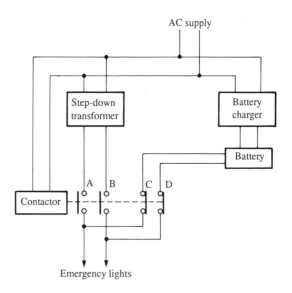

Figure 31 *Maintained system*

Non-maintained system

Here the lighting units remain de-energized until a mains failure occurs, at which time they are illuminated by a battery supply (Figure 32).

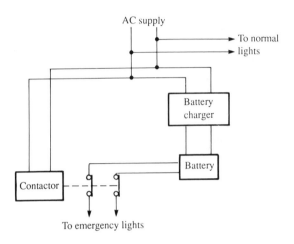

Figure 32 *Non-maintained system*

It should be noted that modern systems use electronic means to provide the change-over from mains to battery supply. The contactor method, however, serves to illustrate the principle of operation.

Security and fire alarm systems

As with emergency lighting, modern security and fire alarm systems are electronically controlled, and it is not the brief of this book to investigate electronic circuitry. However, as with the previous section, the basic principle of operation can be shown by electro-mechanical means.

Both security and fire alarm systems are basically the same in that various sensors are wired to a control panel, which in turn will activate an alarm in the event of sensor detection (Figures 33 and 34). Some modern panels have the facility for incorporating both systems in the same enclosure.

The principles of operation are as follows.

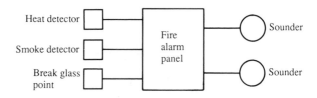

Figure 33   *Block diagram for fire alarm system*

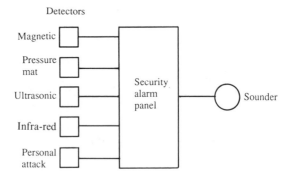

Figure 34   *Block diagram for security alarm system*

Open circuit system

In this system the call points (sensors, detectors etc.) are wired in parallel such that the operation of any one will give supply to the relay RA and the sounder via the reset button (Figure 35). N/O contacts RA1 will then close, holding on the relay and keeping to the sounder via these contacts. This hold-on facility is most important as it ensures that the sounder is not interrupted if an attempt is made to return the activated call point to its original off position.

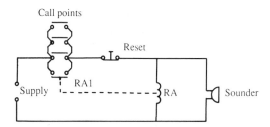

Figure 35 *Open circuit*

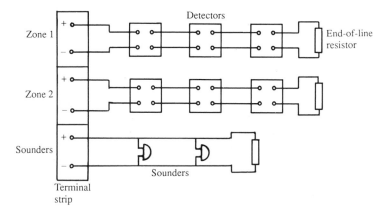

Figure 36 *Fire alarm system*

Fire alarm systems are usually wired on an open circuit basis, with a two-wire system looped from one detector to the next, terminating across an end-of-line resistor (EOLR). This provides a circuit cable monitoring facility; the EOLR is of sufficiently high value to prevent operation of the alarm. Figure 36 shows a typical connection diagram.

Closed circuit system

This system has the call points wired in series, and the operation of the reset button energizes relay RA (Figure 37). N/O contacts RA1 close and N/C contacts RA2 open, the relay RA remaining energized via contacts RA1 when the reset button is released. The alarm sounder mute switch is then closed, and the whole system is now set up.

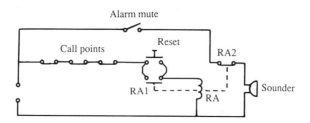

Figure 37 *Closed circuit*

An interruption of the supply to the relay RA, by operation of any call point, will de-energize the relay, open RA1 and close RA2, thus actuating the alarm sounder. The system can only be cancelled and reset by use of the reset button.

The closed circuit system is quite popular, as it is self-monitoring in that any malfunction of the relay or break in the call point wiring will cause operation of the system as if a call point had been activated.

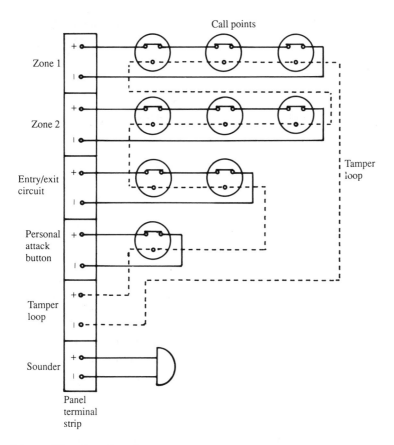

Figure 38 *Security alarm system*

Intruder alarm systems tend, in the main, to be based on the closed circuit type. Figure 38 shows the connection diagram for a simple two-zone system with tamper loop and personal attack button.

The tamper loop is simply a continuous conductor wired to a terminal in each detector in the system. It is continuously monitored irrespective of whether the alarm system is switched on or off, and if interrupted will cause immediate operation of the alarm.

41

The entry/exit circuit is usually confined to the front and/or back doors. The facility exists to alter the time delay between setting the system and exiting, and between entering and switching the system off. This adjustment is made inside the control panel.

All security and fire alarm systems should have battery back-up with charging facilities.

Call systems

Once again these fall into different categories, such as telephone systems and page and bleeper systems. However, the nurse-call variety which uses push-buttons and lamp indication is probably the most popular.

With this type, each room is equipped with a call button of some description, a patient's reassurance light, and a cancel button. Outside each room is an indicator light, and at strategic points in the building there are zone buzzers. Centrally located is a display panel which incorporates a buzzer and an indication of which room is calling.

Figure 39 illustrates, in a simple form, the principle of operation of such a system. This system should by now be quite familiar to the reader; it is simply another variation of the hold-on circuit. Any patient pushing a call button energizes his/her corresponding relay in the main control panel, which is held on by a pair of N/O contacts. At the same time the reassurance, the room and the panel lights 1, 2 and 3 are all illuminated. The zone and panel buzzers are energized via the relay's other pair of N/O contacts.

It is usual to locate the cancel button only in the patient's room, as this ensures that staff visit the patient in question.

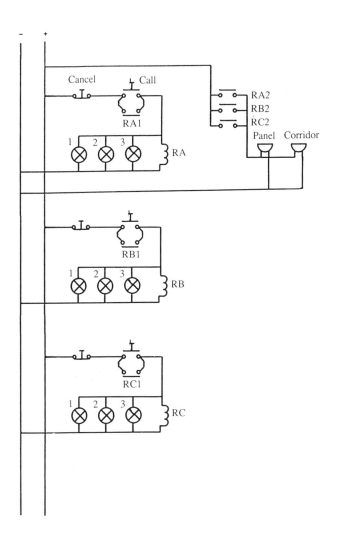

Figure 39 *Nurse-call system*

Motor starter circuits

No book on wiring systems would be complete without reference to control circuits for motor starters. Here we will look at direct-on-line (DOL) and star-delta (SD) starters. Once more, the good old hold-on circuit is employed in both types.

Direct-on-line starter

Both single-phase and three-phase types use the same circuit, as illustrated in Figures 40a and b. For Figure 40b it is important to note that if a 240 V coil were to be used instead of a 415 V type, the coil connections would require a neutral conductor in the starter.

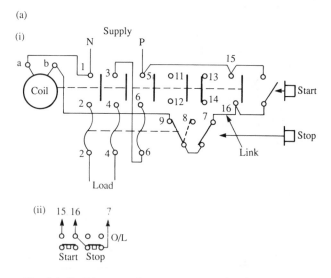

Figure 40 (a) (i) *Direct-on-line starter, single phase.*
(ii) *Connections for remote push-button (start/stop) control: omit link and connect as shown*

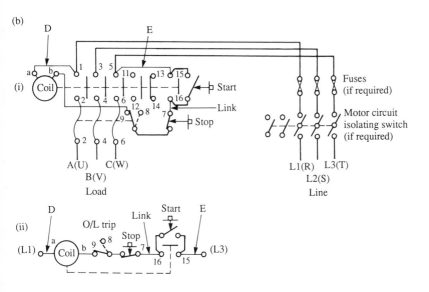

Figure 40 (b) (i) *Direct-on-line starter, three phase.* (ii) *Schematic diagram. Control circuit supply: for phase to phase, connect as shown; for phase to neutral, omit connection D and connect neutral to terminal a; for separate supply, omit D and E, and connect separate coil supply to terminals a and 15*

Star-delta starter

Figures 41a and b show the wiring and schematic diagrams for a star-delta starter. This is clearly a more complicated system than the DOL type. Nevertheless, reference to the schematic Figure 41b will indicate how the system functions:

1   When the start is pushed, supply is given to the star contactor Ⓐ and electronic timer ET, from L1 to L3, and hence all contacts

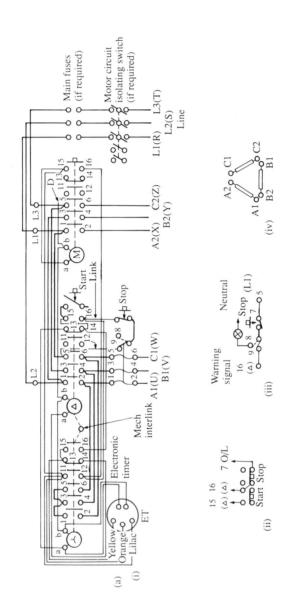

Figure 41 (a) (i) *Automatic star-delta starter.* (ii) *Connections for remote push-button (start/stop) control: omit link and connect as shown.* (iii) *Connections for trip warning.* (iv) *Motor windings: connect to appropriate terminals on starter*

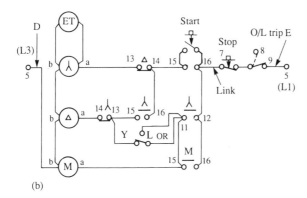

(b)

Figure 41 (b) *Schematic diagram*
Control circuit supply: for phase to phase, connect as shown; for phase to neutral, omit connection D and connect neutral to terminal b; for separate supply, omit D and E, and connect separate coil supply to terminals b and 9
Connections for remote pilot switch control: remove connection 14 to 15 on delta contactor; connect between 14 and 16 on M contactor to terminal 14 on delta contactor; connect pilot switch in place of connection E

marked ⋏ will operate. Supply to Ⓐ and ET is maintained, after the start is released, via ⋏ 11 and 12, timer contacts OR-Y, and ⋏ 15 and 16. The main contactor Ⓜ is also energized via ⋏ 11 and 12 and M15, and thereafter maintained via its own contacts M15 and 16.

2   The motor has of course started. After a predetermined time delay ET operates and its contacts OR-Y change to OR-L. This cuts off supply to Ⓐ and ET. All ⋏ contacts return to normal, and ET resets OR-L to OR-Y.

3   Supply is now given to the delta contactor Ⓐ via M15 and 16, OR-Y, and ⋏ 13 and 14. Delta contacts Δ 13 and 14 open and prevent further energization of the star contactor.

The reader will notice that the line and load terminal markings in Figure 41a show letters in brackets; these are the continental equivalents.

Central heating systems

It would be impossible in such a small book to deal with the vast number of modern systems and variations that are currently available. We will therefore take a look at the two most basic arrangements: the pumped central heating (CH) and gravity-fed hot water (HW) system, and the fully pumped system with mid-position valve. It must be remembered that, whatever the system, it is imperative that the wiring installer has a knowledge of the function of the system in order for him/her to do a competent job.

Pumped CH and gravity HW

This system comprises a boiler with its own thermostat to regulate the water temperature, a pump, a hot water storage tank, a room thermostat, and some form of timed programmer. The water for the HW, i.e. the taps etc., is separate from the CH water, but the boiler heats both systems.

Figure 42 shows such a system. From the diagram it might appear that when the requirement for HW is switched off at the programmer, the CH cannot be called for as the boiler has lost its feed. In fact, such programmers have a mechanical linkage between the switches: HW is allowed without CH, but selection of CH automatically selects HW also.

Note the little heating element in the room thermostat; this is known as an accelerator. Its purpose is to increase the sensitivity of the thermostat; manufacturers claim that it increases the accuracy of the unit to within one degree Celsius. The inclusion of an accelerator (if required) does mean an extra conductor for connection to neutral.

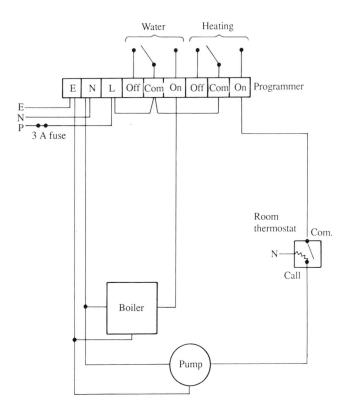

Figure 42 *Gravity primary and pumped heating*

Fully pumped system

Two additional items are required for this system: a cylinder thermostat and a mid-position valve. In this system HW and CH can be selected independently. The mid-position valve has three ports: a motor will drive the valve to either HW only, CH only, or HW and

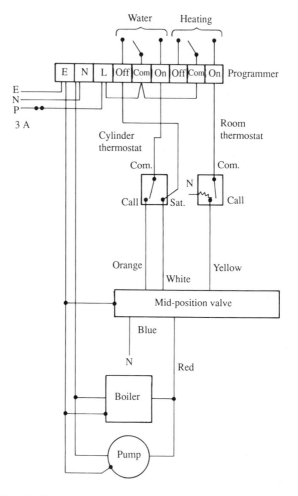

Figure 43 *Fully pumped system*

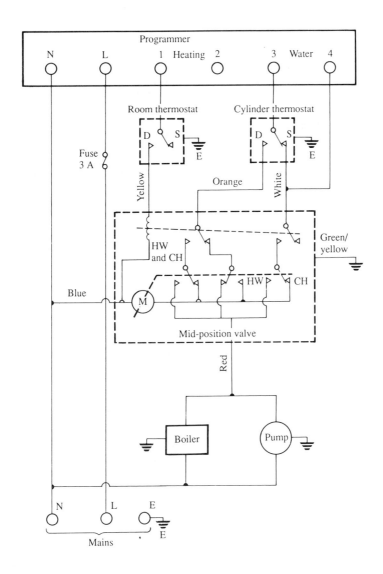

Figure 44 *Internal connections of mid-position valve*

CH combined. With this system the boiler and pump always work together. Figure 43 illustrates the system, and Figure 44 shows the internal connections of a mid-position valve.

Some difficulties may be experienced in wiring when the component parts of the system are produced by different manufacturers. In this case it is probably best to draw one's own wiring diagram from the various details available.

Extra low voltage lighting

These systems, incorrectly referred to as low voltage lighting (low voltage is 50 V to 1000 V AC), operate at 12 V AC. They employ tungsten halogen dichroic lamps, which have a very high performance in comparison with 240 V halogen lamps. For example, a 50 W dichroic lamp has the same light intensity as a 150 W PAR lamp.

Extra low voltage (ELV) lighting is becoming very popular, especially for display purposes. There is very little heat emission, the colour rendering is excellent, and energy consumption is very low.

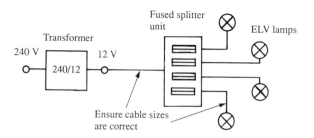

Figure 45

The 12 V AC to supply the lamps is derived from a 240 V/12 V transformer specially designed to cater for the high starting surges, and only these types should be used. The voltage at each lamp is critical: 0.7 V overvoltage can cause premature ageing of the lamp, and 0.7 V undervoltage will reduce the light output by 30 per cent. Hence variation in voltage must be avoided.

To achieve this, leads and cables must be kept as short as possible, and the correct size used to avoid excessive voltage drop. When several lamps are to be run from one transformer, it is advisable to use a fused splitter unit rather than to wire them in a parallel chain (Figure 45).

It is important to remember that, for example, a 50 W ELV lamp will draw 4.17 A from the 12 V secondary of the transformer ($I = P/V$). Although a 1.0 mm cable will carry the current, the voltage drop for only 3 m of this cable will be 0.55 V.

3 Testing and test instruments

Measurement of electrical quantities

As the reader will know, the basic electrical quantities which need to be measured in the world of the installation electrician are current, voltage and resistance. The units of these quantities are the ampere, the volt and the ohm, respectively.

Paradoxically, however, the range and complexity of the instruments available to measure these three fundamental quantities is enormous. So where does one start in order to make a choice of the most suitable types?

Let us look at the range of quantities that an electrician is likely to encounter in the normal practice of his/her profession. If we take the sequence of the more commonly used tests prescribed by the IEE Wiring Regulations, and assign typical values to them, we can at least provide a basis for the choice of the most suitable instruments. It will be seen from the accompanying table that all that is required is an ohmmeter of one sort or another, a residual current device (RCD) tester and an instrument for measuring prospective short circuit current.

	Test	Range	Type of instrument
1	Continuity of ring final conductors	0.05 to 0.8 ohms	Low reading ohmmeter
2	Continuity of protective conductors	2 to 0.005 ohms or less	Low reading ohmmeter
3	Earth electrode resistance	Any value over about 3 or 4 ohms	Special ohmmeter
4	Insulation resistance	Infinity to less than 1 megohm	High reading ohmmeter
5	Polarity	None	Ohmmeter, bell etc.
6	Earth fault loop impedance	0 to 2000 ohms	Special ohmmeter
7	Operation of RCD	5 to 500 mA	Special instrument
8	Prospective short circuit current	2 A to 20 kA	Special instrument

Selection of test instruments

It is clearly most sensible to purchase instruments from one of the established manufacturers rather than to attempt to save money by buying cheaper, lesser known brands. Also, as the instruments used in the world of installation are bound to be subjected to harsh treatment, a robust construction is all-important.

Many of the well known instrument companies provide a dual facility in one instrument, e.g. prospective S/C current and loop impedance, or insulation resistance and continuity. Hence it is likely that only three or four instruments would be needed, together with an approved test lamp (a subject to be dealt with in the next section).

Now, which type to choose, analogue or digital? There are merits in both varieties, and the choice should not be determined just by expense or a reluctance to use 'new-fangled electronic gadgetry'! Accuracy, ease of use and robustness, together with personal preference, are the all-important considerations.

Approved test lamps and indicators

Search your tool boxes; find, with little difficulty one would suspect, your 'neon screwdriver' or 'testascope'; locate a very deep pond; and drop it in!

Imagine actually allowing electric current at low voltage (50 to 1000 V AC) to pass through one's body in order to activate a test lamp! It only takes around 10 to 15 mA to cause a severe electric shock, and 50 mA (1/20th of an ampere) to kill.

Apart from the fact that such a device will register any voltage from about 5 V upwards, the safety of the user depends entirely on the integrity of the current limiting resistor in the unit. An electrician received a considerable shock when using such an instrument after his apprentice had dropped it in a sink of water, simply wiped it dry and replaced it in the tool box. The water had seeped into the device and shorted out the resistor.

An approved test lamp should be of similar construction to that shown in Figure 46.

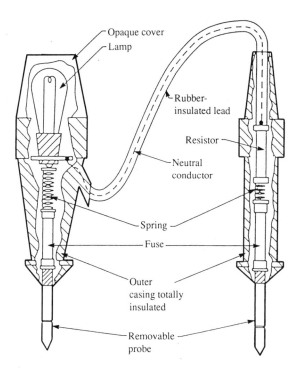

Figure 46 *Approved test lamp*

Accidental RCD operation

It has long been the practice when using a test lamp to probe between phase and earth for indication of a live supply on the phase terminal. However, this can now present a problem where RCDs exist in the circuit, as of course the test is applying a deliberate phase to earth fault.

Some test lamps have LED indicators, and the internal circuitry of such test lamps limits the current to earth to a level below that at which the RCD will operate. The same limiting effect applies to multimeters. However, it is always best to check that the testing device will have no effect on RCDs.

Calibration, zeroing and care of instruments

Precise calibration of instruments is usually well outside the province of the electrician, and would normally be carried out by the manufacturer or a local service representative. A check, however, can be made by the user to determine whether calibration is necessary by comparing readings with an instrument known to be accurate, or by measurement of known values of voltage, resistance etc.

It may be the case that readings are incorrect simply because the instrument is not zeroed before use, or because the internal battery needs replacing. Most modern instruments have battery condition indication, and of course this should never be ignored.

Always adjust any selection switches to the off position after testing. Too many instruments fuses are blown when, for example, a multimeter is inadvertently left on the ohms range and then used to check for mains voltage.

The following set procedure may seem rather basic but should ensure trouble-free testing:

1 Check test leads for obvious defects.
2 Zero the instrument.
3 Select the correct range for the values anticipated. If in doubt, choose the highest range and gradually drop down.
4 Make a record of the test results, if necessary.
5 When a zero reading is expected and occurs (or, in the case of insulation resistance, an infinite reading), make a quick check on the test leads just to ensure that they are not open circuited.
6 Return switches/selectors to the off position.
7 Replace instrument and leads in carrying case.

Continuity of ring final conductors

The requirement of this test is that each conductor of the ring is continuous. It is, however, not sufficient to simply connect an ohmmeter, a bell etc. to the ends of each conductor and obtain a reading or a sound.

So what is wrong with this procedure? A problem arises if an interconnection exists between sockets on the ring, and there is a break in the ring beyond that interconnection. From Figure 47 it will be seen that a simple resistance or bell test will indicate continuity via the interconnection. However, owing to the break, sockets 4 to 11 are supplied by the spur from socket 12 – not a healthy situation. So how can one test to identify interconnections?

There are at present three methods of conducting such a test. Two are based on the principle that resistance changes with a change in length or cross-sectional area; the other relies on the fact that the resistance measured across any diameter of a circular loop of conductor is the same. Let us now consider the first two.

Method 1

If we were to take a length of conductor XYZ and measure the resistance between its ends (Figure 48), then double it over at Y, join X and Z, and measure the resistance between XZ and Y (Figure 49), we would find that the value was approximately 1/4 of the original. This is because the length of the conductor is halved and hence so is the resistance, and the cross-sectional area is doubled and so the resistance is halved again.

In order to apply this principle to a ring final circuit, it is necessary to know the position of the socket nearest the mid-point of the ring.

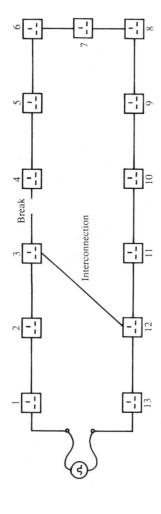

Figure 47 *Ring circuit with interconnection*

Figure 48

$$R_2 \simeq \tfrac{1}{4} R_1$$

Figure 49

The test procedure is then as follows for each of the conductors of the ring:

1 Measure the resistance of the ring conductor under test between its ends before completing the ring in the fuse board. Record this value, say R_1.
2 Complete the ring.
3 Using long test leads, measure between the completed ends and the corresponding terminal at the socket nearest the mid-point of the ring. Record this value, say R_2. (The completed ends correspond to point XZ in Figure 49, and the mid-point to Y.)
4 Measure the resistance of the test leads, say R_3, and subtract this value from R_2, i.e. $R_2 - R_3 = R_4$ say.
5 A comparison between R_1 and R_4 should reveal, if the ring is healthy, that R_4 is approximately $1/4$ of R_1.

Method 2

The second method tests two ring circuit conductors at once, and is based on the following.

Take two conductors XYZ and ABC and measure their resistances (Figure 50). Then double them both over, join the ends XZ and AC and the mid-points YB, and measure the resistance between XZ and AC (Figure 51). This value should be $1/4$ of that for XYZ plus $1/4$ of that for ABC.

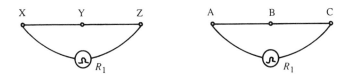

Figure 50

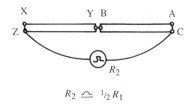

Figure 51

If both conductors are of the same length and cross-sectional area, the resultant value would be 1/2 of that for either of the original resistances.

Applied to a ring final circuit, the test procedure is as follows:

1   Measure the resistance of both phase and neutral conductors before completion of the ring. They should both be the same value, say R_1.
2   Complete the ring for both conductors, and bridge together phase and neutral at the mid-point socket (this corresponds to point YB in Figure 51). Now measure between the completed phase and neutral ends in the fuse board (points XZ and AC in Figure 51). Record this value, say R_2.
3   R_2 should be, for a healthy ring, approximately 1/2 of R_1 for either phase or neutral conductor. When testing the continuity of a circuit protective conductor (CPC) which is a different size from either phase or neutral, the resulting value R_2 should be 1/4 of R_1 for phase or neutral plus 1/4 of R_1 for the CPC.

Method 3

The third method is based on the measurement of resistance at any point across the diameter of a circular loop of conductor (Figure 52).

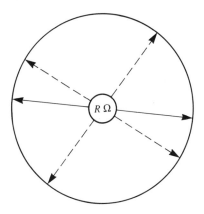

Figure 52

As long as the measurement is made across the diameter of the ring, all values will be the same. The loop of conductor is formed by crossing over and joining the ends of the ring circuit conductors at the fuse board. The test is conducted as follows:

1 Identify both 'legs' of the ring.
2 Join one *phase* and one *neutral* conductor of opposite legs of the ring.
3 Obtain a resistance reading between the other *phase* and *neutral* (Figure 53). (A record of this value is not important.)
4 Join these last two conductors (Figure 54).
5 Measure the resistance value between P and N at each socket on the ring. All values should be the same.

The basic principle of this method is that the resistance measured between any two points, equidistant around a closed loop of conductor, will be the same.

Such a loop is formed by the phase and neutral conductors of a ring dual circuit (Figure 55).

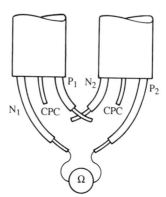

Figure 53

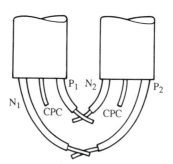

Figure 54

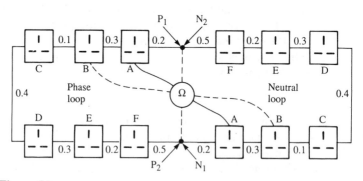

Figure 55

Let the resistance of the conductors be as shown.

R measured between P and N on socket A will be:

$$\frac{0.2+0.5+0.2+0.3+0.4+0.1+0.3}{2}=\frac{2}{2}=1 \text{ ohm}$$

R measured between P and N at socket B will be:

$$\frac{0.3+0.2+0.5+0.2+0.3+0.4+0.1}{2}=\frac{2}{2}=1 \text{ ohm}$$

Hence all sockets on the ring will give a reading of 1 ohm between P and N.

If there were a break in the ring in, say, the neutral conductor, all measurements would have been 2 ohms, incorrectly indicating to the tester that the ring was continuous. Hence step 3 in the test procedure which at least indicates that there is a continuous P–N loop, even if an interconnection exists. Figure 56 shows a healthy ring with interconnection.

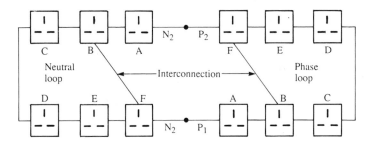

Figure 56 *Healthy ring with interconnection. Sockets A, B and F will give identical readings, C, E and D will not*

The same effect occurs if either or both conductors of the ring are broken beyond the interconnection.

A break before the interconnection would be picked up at step 3 of the test.

Continuity of protective conductors

In this instance we are talking about the CPCs of radial circuits and, of course, main equipotential and supplementary bonding conductors.

As the resistance of copper conductor is known, it is not too difficult to establish whether a conductor of a certain size is continuous throughout its length. For example, a 10 m length of 10.0 mm² main equipotential bonding conductor should give a reading of approximately $10 \times 1.83/1000 = 0.0183$ ohms. The accompanying table gives the resistances of copper conductors in milliohms per metre (mΩ/m) at 20 °C.

Conductor size (mm²)	Resistance (mΩ/m)
1.0	18.1
1.5	12.1
2.5	7.41
4.0	4.61
6.0	3.08
10.0	1.83
16.0	1.15
25.0	0.727
35.0	0.524

Hence it is quite easy to establish the integrity of a length of protective conductor.

It has been known for certain so-called electrical contractors (who wear spurs and ride horses) to use a short length of the correct size of main bonding conductor at each end of a run, and then, for the sake of economy, to join these ends below floor level to a smaller size of conductor!

Earth electrode resistance

If we were to drive an electrode into the earth and then take measurements of resistance between this electrode and another, driven in at increasing distances from the first, we would find that the resistance increased in value for about 2.5 to 3 metres. After this, the values would stabilize and no further increase in resistance would be noted (Figure 57). The final steady value measured is the earth electrode resistance.

The recommended test specified in the IEE Wiring Regulations requires an extra electrode (Figure 58). The test is based on the principle of the potential divider (Figure 59). The auxiliary electrode

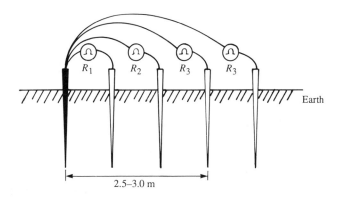

Figure 57

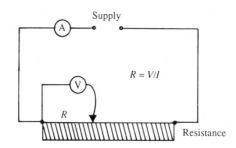

Figure 58

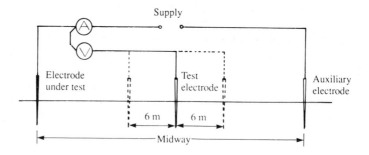

Figure 59

should be placed far enough away to ensure that the test electrode positions do not overlap into any resistance areas. The test may be carried out using a separate voltmeter and ammeter and then calculating the resistance from Ohm's law, i.e. $R = V/I$. However, it is more usual to employ a specifically designed instrument with a direct readout in ohms. As shown in Figure 58, three readings are required; their average is taken as the final value. Naturally this cannot be exact; soil conditions vary widely between areas and seasons, and values are also affected by the size and type of electrode used.

Insulation resistance

This is so important! Damage to the insulation of conductors puts not only persons and livestock at risk of shock, but premises at risk of fire. The resistance between conductors along their length should be very high – of the order of millions of ohms.

It should be pointed out that insulation resistance cannot be measured effectively with a multimeter. The insulation must be stressed to twice the supply voltage (but the test voltage need not exceed 500 V). Multimeters cannot achieve these voltages.

As the resistance of insulation between conductors is in fact countless resistances in parallel, it follows that the longer the conductors, the more parallel resistances there are, and hence the lower the overall insulation resistance. Add to this the fact that installation circuits are also wired in parallel, and one can see that for very large installations an insulation resistance test at the intake position may show a low value, which is not actually due to bad insulation.

In order to overcome this problem, the IEE Wiring Regulations permit such installations to be broken down into smaller units of not less than 50 outlets.

The tests for insulation resistance are as follows.

Poles to earth

1 Isolate supply.
2 Ensure that all protective devices are in place and all switches are closed.
3 Link all poles of the supply together.
4 Test between the linked poles and earth. The reading should not be less than 1 megohm.

Between poles

1   As previous test.
2   As previous test.
3   Remove all accessories and lamps etc.
4   Test between poles (i.e. P to N, or R to Y, R to B, Y to B, and each pole in turn to N). The reading should not be less than 1 megohm.

Tests on disconnected equipment should give values according to the relevant British Standard for that equipment. However, if that information is not available, the value should not be less than 0.5 megohms.

It is worth mentioning that the modern installation is likely to house some electronic devices, and great care should be taken to disconnect such equipment before testing. A test at 500 V can have an adverse effect on timers, programmers, dimmer switches etc.

Polarity

The IEE Wiring Regulations require that all fuses and single-pole devices such as single-pole MCBs and switches are connected in the phase conductor only.

They further recommend that Edison screw lamp holders have the centre contact connected to the phase conductor, and that socket outlets be connected correctly.

Ring final circuits

If method 3 (described earlier) was used to test for ring circuit continuity, phase–earth and neutral–earth reversals would show up

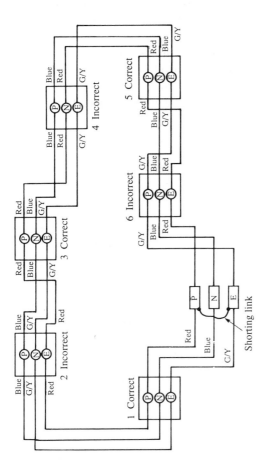

Figure 60 *Polarity on ring final circuit. Sockets 1, 3 and 5 will show as correct; hence it is necessary also to test between phase and neutral with link between main phase and neutral*

at the sockets at which they occur, and would naturally be corrected at that stage. Hence it is only necessary to check for phase–neutral reversals. This can be achieved quite simply by linking the completed ends of the phase and earth rings at the fuse board and then testing between phase and earth at each socket. A zero reading will show reversed polarity.

If methods 1 or 2 were used and the mid-point socket was of correct polarity, reversals elsewhere in the ring would not be detected at that stage. In this case two tests are needed. The first is as just described. The second is to link phase and neutral at the fuse board and test between P and N at each socket; once again a zero reading will indicate a reversal (Figure 60).

Radial circuits

Probably the easiest method here is to test between sockets at corresponding terminals (Figure 61). This of course may involve long test leads.

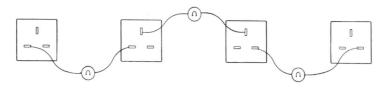

Figure 61   *Polarity test on radial circuit*

Lighting circuits

Short out between phase and earth at the fuse board and test between P and E at each light switch (Figure 62).

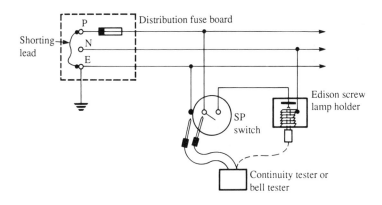

Figure 62 *Polarity test on lighting circuit*

Earth fault loop impedance

Here is a test that has to be carried out on an energized installation, and hence all necessary safety precautions must be taken.

The purpose of the test is to ensure that, in the event of a phase to earth fault, enough current will flow around the earth fault loop path (Figure 63) to operate the protection within a specified time. For socket outlet circuits and bathrooms this time is 0.4 s, and for fixed equipment circuits it is 5 s.

The measured value of loop impedance is then compared with that given in the IEE Wiring Regulations for the particular type and size of protective device. For example, a radial circuit feeding socket outlets and protected by a 20 A rewirable fuse should give a reading, measured from the farthest socket, of 1.8 ohms or less.

Care must be taken to ensure that if the circuit under test is RCD protected, then either the RCD is bypassed or the test instrument is one that is designed not to trip RCDs, as it will be seen that the test is putting a phase–earth fault on the circuit under test.

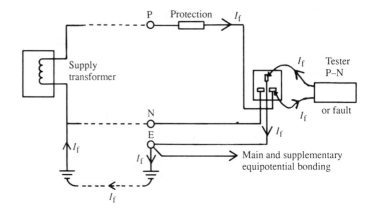

Figure 63 *Earth fault loop path. The earth fault current loop
(phase to earth loop) comprises the following parts, starting at the
point of fault: the circuit protective conductor; the consumer's
earthing terminal and earthing conductor; for TN systems the
metallic return path, or for TT systems the earth return path; the
path through the earthed neutral point of the transformer and the
winding; and the phase conductor from the transformer to the point
of fault. The earth fault loop impedance Z_s at a point distant from
the origin of the installation is measured with the main equipotential
bonding conductors in place*

Operation of RCD

As we have just discussed, phase–earth tests can cause an RCD to
trip. However, an RCD tester is designed to do just that, except that
conditions are controlled, the resultant tripping time being registered.

Most types have a range of settings from 5 to 500 mA and are
simple to use. They are plugged or connected into the circuit under
test, selected to the rating of the RCD in question and operated via a
test button. The time recorded should not exceed that stated by the

manufacturer of the RCD or, if that is not known, should not exceed 200 ms.

If a 30 mA RCD is installed to give added protection against 'direct contact' (i.e. touching live parts), the breaker should trip out in 40 ms at a residual operating current of 150 mA. Most RCDs have the facility to perform this test.

Prospective short circuit current

If we were able to remove all the protective devices in a system and connect a shorting link between phase and neutral conductors, the resulting current that would flow at the point where the link was placed is known as the prospective short circuit current (PSCC). This is all very interesting, you may comment, but why would one need to know this information?

When a short circuit occurs, the device protecting the circuit in question has to open that circuit. If the device is incorrectly selected, a violent and possibly damaging explosion could result. We have probably all experienced the loud bang and resulting blackened and molten copper splattered fuse holders that often occur with rewirable fuses.

Each protective device has a rated breaking capacity, which indicates the level of PSCC that it can interrupt without arcing or scattering of hot particles or damage to associated conductors. For example, rewirable fuses to BS 3036 ranging between 5 A and 200 A have breaking capacities ranging from 1000 A to 12 000 A, and HRC BS88 types from 10 000 A to 80 000 A. MCBs are designed for one fault level irrespective of size, and have ranges of 3000 A (M3), 6000 A (M6), 9000 A (M9) and 16 000 A (M16).

So, it is important to ensure that the breaking capacity of a protective device is capable of interrupting at least the PSCC at the point at which it is installed; hence the need to measure this current.

The type of instrument most commonly used is a dual device, the other part of which is usually designed for earth fault loop impedance testing. It is simply connected or plugged into the circuit close to the protection; a button is pushed and the PSCC value is read.

It is not always necessary to test at every point at which a protective device is installed if the breaking capacity of the lowest rated device in the circuit is greater than the PSCC at the origin. In Figure 64, clearly if the breaking capacity of the 20 A protective device at Y is greater than the PSCC at X, then there is no need to test at the other DBs as the protection there will have even greater breaking capacities than the 20 A device.

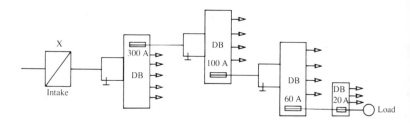

Figure 64

4 *Fault finding*

Starting this chapter has been as difficult as finding a fault on a complex system. The old adage of 'start at the beginning and finish at the end' is all right for storytellers or authors of technical books, but is not always sensible for fault finders! Perhaps it would be wise to begin by determining what exactly an electrical fault is.

A fault is probably best defined as a disturbance in an electrical system of such magnitude as to cause a malfunction of that system. It must be remembered, of course, that such disturbances may be the secondary effects of (a) mechanical damage or (b) equipment failure. Actual electrical faults, or should one say faults caused by 'electricity', are rare and are confined in the main to bad design and/or installation or deterioration and ageing.

However, all of this still does little to indicate where to begin looking for a fault. Only experience will allow one to pinpoint the exact seat of a breakdown, but a logical approach to fault finding will save a great deal of time and frustration.

Signs and symptoms

Always look for tell-tale signs that may indicate what kind of fault has caused the system breakdown. If possible, ask persons present to describe the events that led up to the fault. For example, inspection of a rewirable fuse may reveal one of the conditions shown in Figures 65

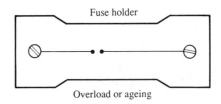

Figure 65

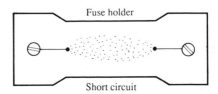

Figure 66

and 66. The IEE Wiring Regulations define overload and short circuit current as follows:

Overload current   An overcurrent occurring in a circuit which is electrically sound.
Short circuit current   An overcurrent resulting from a fault of negligible impedance between live conductors having a difference in potential under normal operating conditions.

It is often likely that other personnel will have replaced fuses or reset MCBs to no avail before you, the tester, are called in. In these circumstances they will be in a position to indicate that protective devices trip immediately or after a delay. This at least will give some clue to the type of fault.

Let us now consider typical faults on some of the systems already discussed in this book, and endeavour to suggest some solutions. Remember, however, that it is impossible to catalogue every fault and its cause that may occur in every system.

Ring and radial socket outlet circuits

Reported fault

'My fuse/MCB blows at odd unspecified times during the day, with no particular appliance plugged in to any particular socket.'

Diagnosis

1 This could be incorrect fusing, but that is unlikely because then the fault would always have been present. It is probably an insulation breakdown.
2 Conduct an insulation resistance test. If the reading is low, then:
3 Go to the centre of the circuit and disconnect the socket; disconnect the ends in the fuse board if the circuit is a ring; and test in both directions.
4 Probably only one side will indicate a fault, so subdivide. Keep testing and subdividing until the faulty cable section or socket is located.

New ring final circuit installations

In Chapter 3 three methods are described for testing the continuity of ring final conductors. The purpose is to locate interconnections in the ring. Should such an interconnection prove to exist, its position in the ring must be located.

Methods 1 and 2 will indicate a fault, but location is achieved by systematically removing sockets to find the interconnection. Method 3 will give an indication of the location of the fault because tests on sockets nearest to the fuse board will give similar readings, whereas those beyond the interconnection will be considerably different.

Radial circuits feeding fixed equipment

Reported fault

'Every time I start my hydraulic press, the fuse blows in the board.'

Diagnosis

1   This is probably a fault on the press motor windings or the starter. A supply cable fault would normally operate the protection without an attempt to start the motor.
2   Check the starter for obvious signs of damage, burning etc. If it seems all right, then:
3   Do an insulation resistance test on the motor windings.

Reported fault

'The MCB that protects my roof extraction fan has tripped out and will not reset even with the fan switch off.'

Diagnosis

This is almost certainly a cable fault. Do an insulation resistance test. If a fault is indicated, start looking!

Cable fault location

The comment 'start looking' in the last example is quite acceptable for shortish, accessible cable runs and where some sign of damage is evident. However, for longer, hidden cable routes, especially underground, visual location is impossible. In these instances special tests must be applied, such as the following.

Murray loop test

This test and its variations are based on the principle of the Wheatstone bridge. It is used for the location of short circuit faults.

In Figure 67, when the variable resistances are adjusted such that the reading on the ammeter is zero, it will be found that

$$A/B = C/D$$

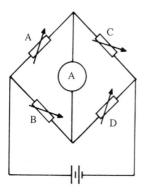

Figure 67

Let us now replace resistors C and D with the cores of a faulty cable (Figure 68). The ratios will be the same, as resistance is proportional to length. A link is placed between a sound core and a faulty core. It will be seen that now C is replaced by $L+(L-X)$, and D by X, so that

$$A/B=[L+(L-X)]/X$$

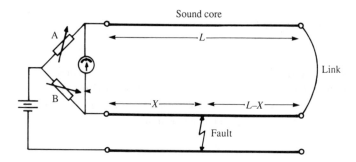

Figure 68

Hence

$$A/B=(2L-X)/X$$
$$AX=2LB-BX$$
$$AX+BX=2LB$$
$$X(A+B)=2LB$$

$$X=\frac{2LB}{A+B}$$

A greater accuracy can be achieved if the test is then repeated at the other end of the cable and the mean position of X is taken as the point of fault.

Variable resistors *A* and *B* are usually incorporated in a single bridge instrument, and the resistances are adjusted for a zero reading.

Emergency lighting

There is not a lot to go wrong with these systems other than cable faults and internal failure of the unit. Cable faults are located in the way previously described; however, do *not* conduct insulation resistance tests with the unit still connected, or the internal electronic circuitry may be damaged.

Individual units have an LED that indicates a healthy state of charge. If this light is out, then:

1   Check the mains supply. If it is all right, then:
2   Check the output of the charging unit at the battery terminals. If this is all right, check the battery separately.

If the LED is lit but the emergency light does not function when the mains supply is removed, check the lamps. If these are functioning, the circuitry is probably faulty.

Security and fire alarm systems

Fire alarm systems have to be inspected at regular intervals and hence tend to be very reliable. The most common faults in such systems are false operation of heat or smoke detectors, perhaps owing to a change in the use of the protected area since they were first installed, or a break in a sensor cable. As the continuity of these cables is monitored, any fault will bring up a fault condition on the main panel. Simple continuity tests on cable between sensors should reveal the broken section.

Security systems are somewhat similar in that faults on cables or sensors will bring up an alarm condition on the main panel. Remove the ends of the zone in question and insert a shorting link. If the alarm still activates when switched on, a fault in the internal circuitry is indicated. If not, start at the middle of a zone and work back to the panel, similar to the procedure used in ring/radial circuits, until the faulty cable or sensor is found.

Call systems

The faults on these systems, once again, are broadly similar to those on alarm systems.

Reported fault

'The patient in room 42 pushes the call button but nothing happens.'

Diagnosis

This is either a faulty push-button, a break in the cable, or a defective relay or electronic component. Trial and error is the rule here: check out the simple and obvious first before taking the control panel apart.

Central heating systems

The most difficult faults to locate are those that occur during the commissioning of a newly installed system, and are almost always due to incorrect wiring. Naturally, not all the faults that could occur on all the available systems can be detailed, but there are one or two common mistakes that are worth mentioning:

1 The CH is selected but the HW comes on, and vice versa. This is simply a reversal of the CH and HW feeds in the programmer.
2 On a mid-position valve system, a reversal of the leads to the demand and satisfy terminals in the cylinder thermostat will cause various incorrect sequences depending on which heating options have been chosen.

On one occasion the author installed the control wiring for a CH system, and was perplexed when the room thermostat operated the HW and the cylinder thermostat the CH. Much tearing of hair followed, as the system wiring proved to be correct. Some while later it was discovered that the DIY householder had installed the mid-position valve back-to-front.

With existing systems, faults tend to be confined to equipment failure such as defective programmers, valves and pumps. Once again it must be stressed that fault location on such systems is eased considerably when the system is fully understood, and generally this understanding comes with experience.

Motor starter circuits

The usual cable faults we have already discussed must of course be considered, but faults in the starter itself are usually confined to coil failure or contact deterioration.

Coil failure

If this should occur the motor will not start, for the obvious reason that the control circuit is defective.

Reported fault

'Everything was all right this morning, the motor was running; but this afternoon everything just stopped, and will not start up again.'

Diagnosis

Could be operation of circuit protection, coil failure or overload operation due to a motor fault.

1 Check the overloads. If they have operated, do an insulation resistance test on the motor windings. If not, then:
2 Check the circuit protection. If it is unaffected, the fault is probably coil failure. Verify and replace.

Contact deterioration

Here is one of the few genuine faults caused by the effects of 'electricity'. Over time, constant operation with its attendant arcing will burn and pit the surface of the contacts. This can lead to poor connection between the contact faces, or even complete destruction of the contacts.

Reported fault

'We shut down the conveyor for lunch, and when we try to start it up again it only works if we keep the start button in.'

Diagnosis

1 This is clearly a 'hold-on' circuit fault. The coil must be OK or

the conveyor would not work, so refer to the starter circuit diagram (if there is one, it is usually on the lid); then:

2   Check the continuity across the hold-on contacts with the contactor depressed manually. If open circuit is found, strip the contactor and investigate.

Protection faults

Other causes of motor circuit failure often result from incorrect protection – fuses, MCBs or overloads – or incorrect reconnection after motor replacement.

Motors may take as much as eight times their normal rated current on starting. If the protection is of the wrong size or type, it will operate.

Fuses   Use a motor rated fuse, usually BS 88. These are easily recognizable as the rating has a prefix M, e.g. M20 is a 20 A motor rated fuse. They are designed to cater for high starting currents.

MCBs   Usually, type 1 and type 2 are too sensitive to handle starting currents, so a type 3 is preferred.

Overloads   These are adjustable and their ratings must be suitable for the full load of the motor. Ensure that the overload has been selected to the correct rating.

Note: do *not* attempt to uprate a protective device just because it keeps tripping out. It is likely that the wrong type rather than the wrong size was initially installed.

Motor replacement

If a motor has been diagnosed as being faulty, and is either repaired or replaced, always ensure that the reconnections are correct. A reversal of the connections to the windings of a three-phase star-delta motor could have serious implications for the motor as well as for the operation of the circuit protection.

Figures 69 and 70 show the terminal arrangements for a six-terminal motor and the correct winding connections.

If the connections to any winding are reversed, the magnetic fields will work against each other and a serious overload will occur, especially when the starter changes the motor to the delta configuration. For example, *never* allow A2 to be connected to B2 and A1 to C1.

This must not be confused with the reversal of any two *phases* of the supply when motor rotation needs to be reversed.

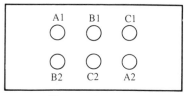

Motor terminal block

Figure 69

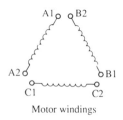

Motor windings

Figure 70

Conclusion – and a cautionary tale

Finally, it must be reiterated that successful and efficient fault finding has its basis in experience. However, for the beginner, use a logical and methodical approach, ask questions of the customer or client, and use some common sense. This in itself will often provide the clues for the diagnosis of system failure.

Once upon a time there was a young electrical apprentice (ahem! who shall remain nameless) who was doing a little wiring job (not private, of course). The customer asked that while the apprentice was there, would he please investigate whether a cracked socket outlet needed changing. Of course he would (thinks: 'Extras!'), and proceeded to undo the socket without isolating. There was a pop and a splash and the ring circuit fuse blew. 'Silly me,' he thought, 'I'll just replace the fuse. I must have shorted something out in the socket.' 'Hold on,' thought he, 'do this properly: never replace a fuse until the circuit has been checked first.'

An insulation resistance test seemed to be the order of the day, so naturally all appliances were removed from the ring and the test was applied. It showed a P to N fault. 'Oh! bother' (or some such expletive) he muttered, 'where do I start?'

Unfortunately the ring circuit was wired in the form of junction boxes in the upstairs floor void with single drops to each socket. Splitting the circuit up meant gaining access to those JBs. Needless to say there were fitted carpets and tongued and grooved boards, and no access traps.

Five hours later, with carpets and floorboards everywhere and a puzzled customer who did not understand such an upheaval to replace a cracked socket, the faulty drop was found. The apprentice ran downstairs but could not immediately find the associated socket outlet. It was there, of course, in a cupboard, and next to it was a fridge with a door like a cupboard door, and of course the fridge was plugged into the socket.

'Heavens above!' and 'Flip me!,' said the apprentice, 'I wish I had asked if there were any appliances I had missed.' Worse than that, 'How do I explain this one away?' Well, needless to say, he did, using the most impressive technical jargon, and vowed never to make the same mistake again.

Solutions

Quiz controller (Chapter 1)

Figure 71 shows the solution. Any contestant pushing his/her button energizes their corresponding relay, which is held on via contacts RA1, RB1 or RC1. Two sets of N/C contacts, located one in each of the other contestants' circuits, will open, rendering those circuits inoperative. The system is returned to normal when the reset button is pushed, de-energizing the held-on relay.

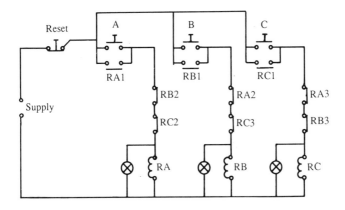

Figure 71 *Quiz controller*

Drawing exercises (Chapter 1)

1 (a) Figure 72.
  (b) Figure 73.
  (c) Figure 74.
  (d) Figure 75.
2 Figure 76.

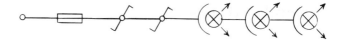

Figure 73

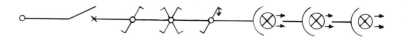

Figure 74

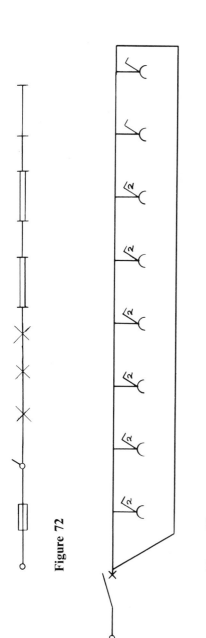

Figure 72

Figure 75

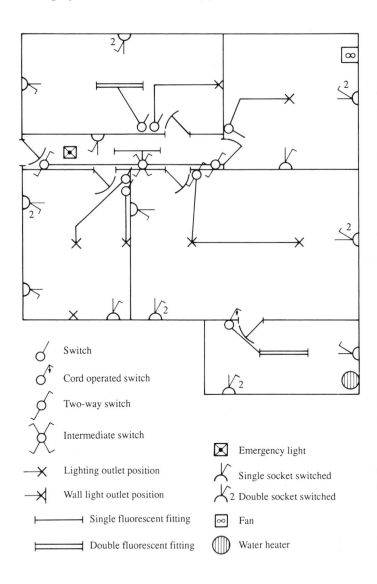

Switch

Cord operated switch

Two-way switch

Intermediate switch

Lighting outlet position

Wall light outlet position

Single fluorescent fitting

Double fluorescent fitting

Emergency light

Single socket switched

2 Double socket switched

Fan

Water heater

Figure 76

94

Index

Index